This book belongs to:

_____

From:

_____

Date:

_____

GOOD NIGHT (NOT RE★LLY)

LET'S COUNT FORW★RD AND B★CKW★RD!

Written by
Nan Evenson

Illustrated by
Karina Matkevych

CKBooks Publishing

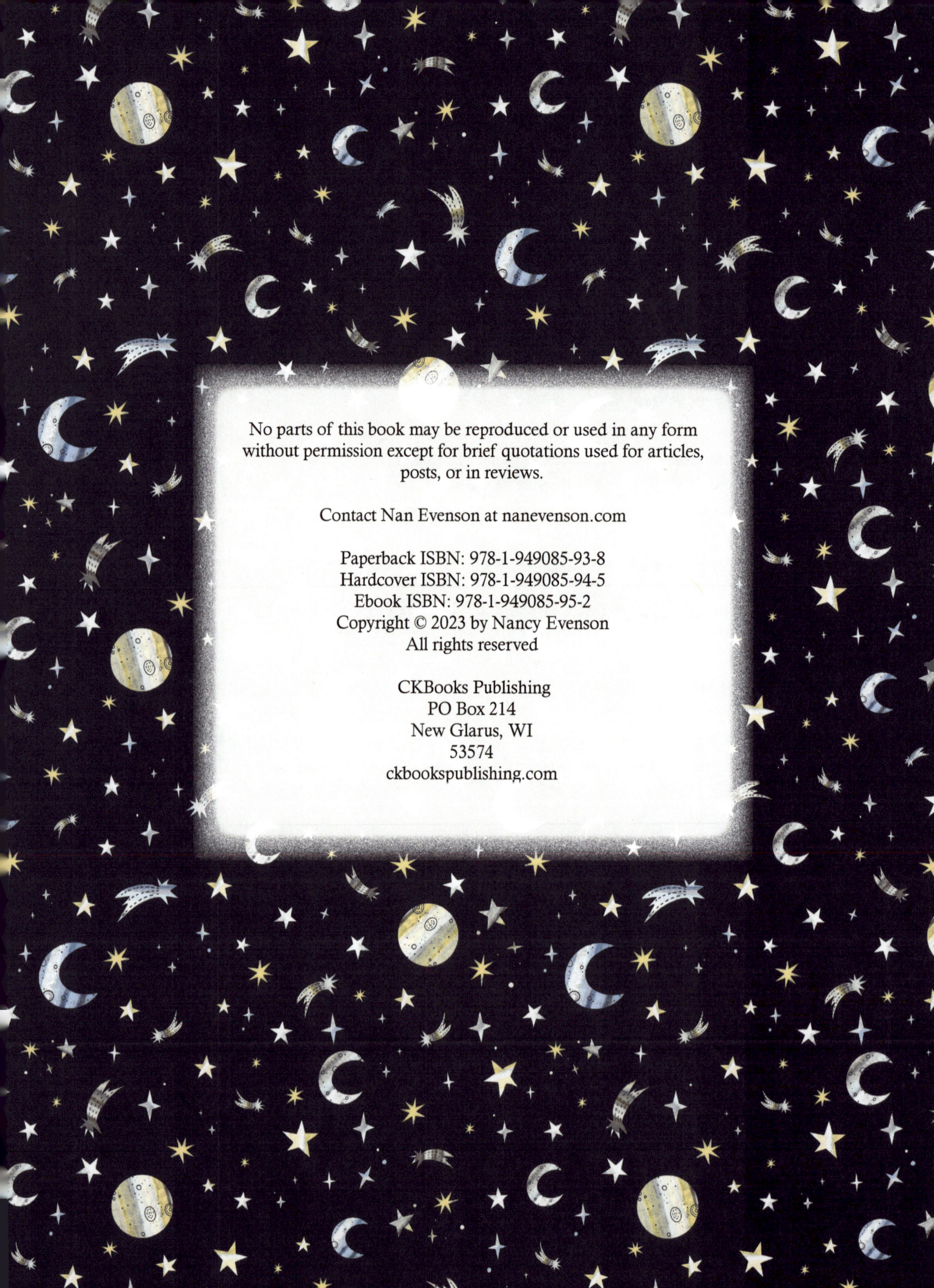

No parts of this book may be reproduced or used in any form without permission except for brief quotations used for articles, posts, or in reviews.

Contact Nan Evenson at nanevenson.com

Paperback ISBN: 978-1-949085-93-8
Hardcover ISBN: 978-1-949085-94-5
Ebook ISBN: 978-1-949085-95-2
Copyright © 2023 by Nancy Evenson
All rights reserved

CKBooks Publishing
PO Box 214
New Glarus, WI
53574
ckbookspublishing.com

For Nadia, Jackson, Zoe, and Marta.

You fill my heart with love.

This is a very SPECIAL good night book!

Because this book is so sweet, it's like  ginormous candy explosion in the playroom.

It's so big and bad, it's like  elephants

are dancing on your bed in a clompy,

stompy elephant way.

It's so noisy, it's like 3 billy goats

are having a meeting in the hall

and are talking very loudly

about unimportant things.

It's so buzzy, it's like 4 little fairies are fluttering around in the bathroom all trying to brush their tiny teeth at once.

It's so delicious, it's like 5 weird broccolis in the kitchen have suddenly turned into yummy, green cupcakes. Wow!

It's so marchy, it's like  gardening tools in the garage are having a parade.

It's so fun, it's like having 7 wacky swings in the backyard.

It's so scary, it's like 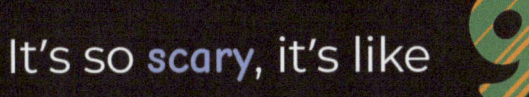...

Wait, wait I thought we were on !

Ah, yes, you ARE a smart child.

It's so mixed-up, it's like 8 BIG mice are playing tag with 9 mini cats in the front yard.

It's so brave, it's like having 10 strong squirrels catch you when you fall.

Hooray! Hooray! What a lucky day!

Let's count BACKWARD now!

10 brave squirrels, 9 mini cats, 8 BIG mice, 7 wacky swings, 6 marchy gardening tools, 5 weird broccolis, 4 buzzy fairies, 3 noisy billy goats, 2 big and bad elephants and 1 ginormous candy explosion.

I thought this was a good night book,

but it's SO hard to go to sleep with

all this hubbub going on.

Well, that's because this is NOT REALLY a good night book.

It's a GOOD MORNING book!

Wake up, kids!

Welcome to this special day!

About the Author

Nan has enjoyed working with youth for over twenty years, helping them with math, reading, and writing. Several of her short stores have been published and two have won awards.
Visit her on her website: nanevenson.com

About the Illustrator

Karina is an extraordinary artist from Ukraine.
Karina is delighted to create beautiful books for smart, little minds.
Visit her on instagram: @matkevych.art

www.ingramcontent.com/pod-product-compliance
Lightning Source LLC
Chambersburg PA
CBHW041530070526
44586CB00002B/29